MEMORIES of GAZOS CREEK
& PIGEON POINT

MEMORIES of GAZOS CREEK & PIGEON POINT

1916 – 1918

by Edward Conant

Glenhaven Press Modesto — 1998

MEMORIES of GAZOS CREEK
& PIGEON POINT
1916 – 1918
by Edward Conant

Edited by: Bessie Conant

Published by:

GLENHAVEN PRESS
2401 E. Orangeburg Ave.
Suite 675 - 109
Modesto, CA 95355-3351

First Printing 1998

Publisher's Cataloging in Publication Data
Conant, Edward
Memories of Gazos Creek & Pigeon Point — 1916-1918

1. History (U.S.) – California—San Mateo Co. Gazos Creek–Pigeon Point, 1916-1918. 2. A woodsman's reminiscences. 3. Gazos Creek sawmills –logging operations –lumber companies. 4. Felling timber –redwoods–woodsmen–skills–spliting redwood. 5. Pigeon Point–wharf–lumber shipped. 6. Genealogy – California – San Mateo Co. – Santa Clara Co. – Santa Cruz Co.

Library of Congress Catalog Card Number: 98-72358
Hard cover: ISBN 0-9637265-1-X

Typeset with PTI LaTeX
Font: ITC Souvenir Light
Photo processing by
Anderson Custom Photo Lab

Cover Art by Studio II

PUBLISHER'S FOREWORD

A WOODSMAN SPEAKS here in his own words — telling of lumber camp life in the days of oxen, skid roads, donkey engines, spring boards, the double-bit axe and the two-man saw used to fell redwoods. The redwoods cut nowadays are saplings compared to the trees cut by Ed Conant. He vividly describes the techniques, labor and the final sound of bringing down one of these mighty monarchs. A second generation woodsman, he describes life growing up in a lumber camp and his boyhood years at Pigeon Point. This book contains history, but it is not a history book. It is a book about people and a life they lived that we can only imagine — about the sights, the sounds, the dangers, the rigors and joys of family, woods and mill — told in the idiom of the woodsman.

MEMORIES of GAZOS CREEK & PIGEON POINT

— TABLE OF CONTENTS —

FOREWORD

LOOKING BACK over the years, especially those of us who are eighty years and older, we seem to have a longing in our hearts for what is called "the good old days of yesterday." They were not really as good as we like to think of them. They could be called the hard days of our life. But also we have to call them good in the sense of family life. When we went to visit we stayed a while as we did not go very often. I recall when Dad went from Pescadero to Stockton, which is about a hundred and twenty-five miles. With a horse and wagon, it took three days, now days about two hours. When night came, wherever you happened to be you just pulled out alongside the road and camped. That was your motel. You built a fire, cooked supper, tied the horse to a post where he could get some grass for his supper. When it came time to crawl into bed, the ground was our Sealy mattress. A kerosene lantern was our light if you could find it. In those days it wasn't called kerosene, it was called coal oil, which name has gone with the years. Many young people never heard the words "coal oil." It will date you if you remember that era of time.

That was my job, to fill the lamps with coal oil and trim the wick so it would not have a high corner and smudge up the lamp shade when you turned up the wick to get a little more light.

Some years later we got what they called aladdin lamps. With this lamp we really came out of the dark. It had what they called a mantle made out of a substance that if you touched it the whole thing would crumble to an ash. This lamp gave a brilliant white light. When you first lit the lamp you had to turn the wick down real low for a few minutes until it warmed up or what we called generated, otherwise it would blacken the mantle. Then you had to turn the light way down so it would burn the spot off the mantle before turning it up again.

— Conant Collection.

Ed Conant in front of the old house where Norman Steele was born. The location is about a half mile south of Gazos Road.

I remember when we were in a hurry for the light we would take a salt shaker and put salt down the chimney and the spot on the mantle would burn off much faster. The reason the black spot came on the mantle was caused by a high spot on the wick or we turned the wick up too high.

You see, in those good old days which we call them, you had to prepare ahead to go to town or for the evenings. For the evenings you had to go out to the chopping block and cut your wood and kindling, go to the spring or to the old hand pump with your buckets and bring in the water, - no inside plumbing. I remember one place we lived I had to go a quarter mile one way to the old well with two buckets to get water.

Sometimes I would get to the well where it had an old hand pump and somebody didn't leave water in a bucket to prime the pump. If you did not prime the pump you could pump on the handle all day long and get no water. To get a bath in those days was something else. You'd build up the fire in the old wood stove, then dump the hot water by the bucketful into the old wash tub, then add some cold water, then crawl into the wash tub which had no room for a person to begin with. Finally you got your bath if you want to call it that. Then you had to take and empty the tub by the bucketfuls outside, then after that, dry the floor of the water that should have been on you but went on the floor instead. So much for that.

These were just a few things that entered my mind before writing about Gazos Creek. I know there have been a number of books and items written on the Gazos and the sawmills and Pigeon Point Lighthouse. These are activities that took place that I would like to go into a little more in detail. The era I would like to write about are the years 1916, 1917, and 1918, the years in which I lived in the Gazos and at Pigeon Point. I am writing this in memory of my father, who started working in sawmills and logging camps in the 1890s. Dad started working when he was fifteen years of age greasing skids. From a skid greaser to a bull team driver for different sawmills.

Our family followed that life for a number of years. My father was a number one logger, also working in the sawmill. I worked with Dad falling timber for a while and he was an all round woodsman. So naturally in my later years I followed the sawmills and logging life, I suppose from just hearing of Dad's experiences and looking at pictures he had taken from those early days in the sawmills. I suppose like-father-like-son is true to a certain degree.

The GAZOS CREEK area as surveyed in 1895 & 1899 by the U.S. Coast and Geodetic Survey. Santa Cruz Quadrangle, 1902 Ed. Courtesy USGS, Menlo Park, CA.

I

THE CAMPBELL LUMBER COMPANY

THE CAMPBELL MILL was two miles up the Gazos from where the Cloverdale Road intersects the Gazos Road. The Campbell sawmill was formerly the Bloom Mill. Bloom operated this mill for five or six years, before shutting it down for the lack of money. In other words, he went broke, as all the others did. The mill laid idle for around three years, when Mr. Curtis Miracle from Campbell, California, a friend of my family for many years, became interested in the mill.

He formed a company of four men and bought the mill and timber. There were over four hundred acres in one parcel of timber, with the option to buy over fourteen acres in another parcel.

The new owners of the former Bloom Mill were Mr. Miracle, Mr. Strom, Mr. Duncan, and Mr. Kramer. In the fall of the year 1916 the company was formed. Most of the new owners were from Campbell, California, so they called the former Bloom Mill the Campbell Redwood Lumber Mill.

The mill had been shut down for around three years and in that time of idleness, lots of the machinery had been taken or stolen out of the mill. So there was much work to be done in preparing the mill for operation. This was in September, 1916, so they had all winter to prepare the mill for operation in the spring of 1917.

All the sawmills in those days were steam operated. The Campbell Mill had a very tall smoke stack. The tall stack was to create a better draft. Steam was a very efficient source of power. They used the waste from the logs and sawdust to fire the boilers to operate the mill. Since the slabs were wet and green and the mill sat right on the bottom of the canyon on the Gazos Creek, the draft was not always so good. So the Campbell Mill installed another smoke stack. There were not too many mills with two smoke stacks.

I recall the summer of 1917 when the mill was operating, the smoke from the burning pit and the smoke stacks (on a day when it was real still, no wind) would just lay in the Gazos in layers, it seemed. The slabs and trimmings and sawdust were more than could be used for firing the boiler, so the excess went out on a chain and into an open pit where it was burned.

Mr. Miracle, an old friend of Dad's, wanted him to come to the Gazos to work. Dad was working felling timber for a large sawmill in Westwood, California. So Dad accepted the job from Mr. Miracle.

Dad and Mother and we four children moved to the Gazos Mill in September, 1916. All our furniture, if you could call it that, came to San Francisco, then was put on the Ocean Shore Railroad and brought to Tunitas. That was the end of the railroad line.

Dad took a team of horses and wagon from the Gazos Mill to Tunitas, which was around eighteen miles, and brought our furniture to Gazos. Freight moved very slowly in those days. Our family was at the mill with no furniture or beds. We had to sleep on the floor of the cabin and ate at the cook house.

Our cabin was on a hill above the commissary and the cook house on the north side of the canyon. There was no road up to our cabin, only a three foot wide trail about 200 yards long and very steep. We had to take all of our belongings up that trail by horse and sled. There were eight cabins on that hill, with eight families.

There were, in the neighborhood, twenty-two or twenty-three cabins plus a cook house, a commissary, a blacksmith shop, a water tank, a barn, a corral, a sawmill, the mill pond, a sawdust and ash burning pit, and a saw filers shed. There were living quarters in the end of the cook house for the cook.

Mother cooked for the loggers and sawmill crew when the cook was ill or away for a few days.

The cook house and the commissary were almost one building, except for about five feet between them. The commissary was about twenty feet long, and the cook house was about fifty feet long with the kitchen and a room for the cook to sleep. It was large enough to feed around one hundred men.

There were two tables with wooden benches down each side, and a walkway down the middle and sides. The tables were made of rough lumber. I ate in the cook house for about a month, until our furniture arrived.

If you never ate at a sawmill cook house in the early nineteen hundreds, you have no idea of what the sanitary conditions were like. There were flies everywhere. The food was good and there was plenty of it. The loggers and the sawmill men worked hard and ate hearty.

I will never forget one cook, he made pies for dessert, and he always put a crust over the top of the pies. After he baked them, he set them on a shelf to cool. By supper time the flies had roosted on the pies by the score. I am not exaggerating a bit. It looked like pepper from fly specs on top of the crust. You would never fail to find a dead fly under the crust.

The dishes we ate from were big, heavy, white porcelain platters. The cups were the same. They weighed more than a cup of coffee today, empty. I still have one of the cups, the sides are three eighths of an inch thick.

There were many mills on the Gazos. Shingle mills were probably the first.

- Burch and Steen Shingle Mill — 1867-1871
- Templeton Mill, later called McKinley Mill — 1871-1873
- Beadle Mill — 1917-1918
- Grover and Littlefield Mill — 1901-1908.
  My uncle worked there in 1905. It was located three miles from Highway One.
- Glenwood Lumber Company taken over by Mr. Bloom in 1905 — known as the Bloom Mill. It was operated for eight years and went broke in 1913.
- The Campbell Lumber Company where I lived, later bought the timber and mill.
- Johansen Mill — two settings:
  - 1. In the head of the Gazos Creek
  - 2. Moved to the other setting - 1920 - up the China Grade
    I worked for Mr. Johansen from 1937-1938.
- California Tie Camp and California Shingle Company, owned by Mr. Cecil Gibson and Mr. Moore — 1930. Mr. Gibson moved the mill to Oil Creek in 1938. I worked for Mr. Cecil Gibson three years — August, 1938 to September, 1940 and again in 1946.

[Facing page.]

TEMPLETON PACIFIC COMPANY 1871-1873. James and David McKinley operated this mill in 1882 and called it the McKinley Mill. It was located across the road from the site of the FORSYTHE AND ALLEN MILL - which was built later, in the 1940s. The McKinley Mill site was a few hundred yards above the Beadle Mill in Gazos.

- Mr. Irwin Dale, a good friend of Dad's and mine, built a Box and Shuck Mill in 1940. Dad helped build the mill. It was located three miles from Highway One. The mill was on the right-hand side of the Gazos Road, south of Gazos Creek. The mill sat on Mr. Humphrey's property. Irwin Dale's Box Mill went broke in less than a year. There was no market for redwood boxes. Today the market is good — hanging baskets are very popular. That was the last mill to be built in the Gazos.

- Forsythe and Allen Mill - 1940 - operated for a short while. The mill was torn down and moved to the Sierra Nevada's. The mill was operated on Butane gas. That mill was located five miles up the Gazos Road from Highway One.

I realize everything was cheaper, so was lumber. In those days, you worked all day, and more, to pay for a pair of overalls, all day for a top shirt and all day for a pair of shoes. Some mills got eighteen to twenty dollars per thousand board feet of lumber — delivered. So you see, by the time you paid for the stumpage, felled it, and got it logged to the mill, then sawed into lumber and hauled to San Francisco, put on the dock, then to the lumber yard; that is a lot of work and handling. So your profit was very small. That is why so many mills went broke, there was nothing left.

Then in the 1920s a Japanese group went into the Gazos to open the Campbell Mill again. They were going to change from the Bull Donkey and bring the logs to the mill pond with a railroad and Shay engine. They had no sawmill experience or logging experience. They spent a lot of money preparing the old skid road for a railroad. Soon they went broke. The old mill never operated again after Campbell Lumber Company shut down in 1918.

[Facing page.]

GROVER & LITTLEFIELD MILL - Joe Gibson 2nd from top left — ca. 1904.

CAZOS. CREEK MILL.

— Conant Collection.

GAZOS MILL owned by BLOOM - 1908 to 1913.

[Facing page.]

GAZOS MILL built by A.W. Rogers and E. Schnabal in 1905. In 1908, BLOOM bought the mill and ran it until 1913. In 1916, the Campbell Lumber Company bought it.

— Conant Collection.

— Harvey Mowry Collection.

GAZOS CREEK MILL - 1920.
Mrs. Mowry with son, Harvey and daughter.

[Facing page.]

IRWIN DALE'S BOX MILL in the Gazos. Built in 1940 by Irwin Dale, Jim Conant and Walter Hohl - on Mr. Humphrey's property. Located three miles from Highway 1. Irwin Dale is shown with wife and sons.

— Conant Collection.

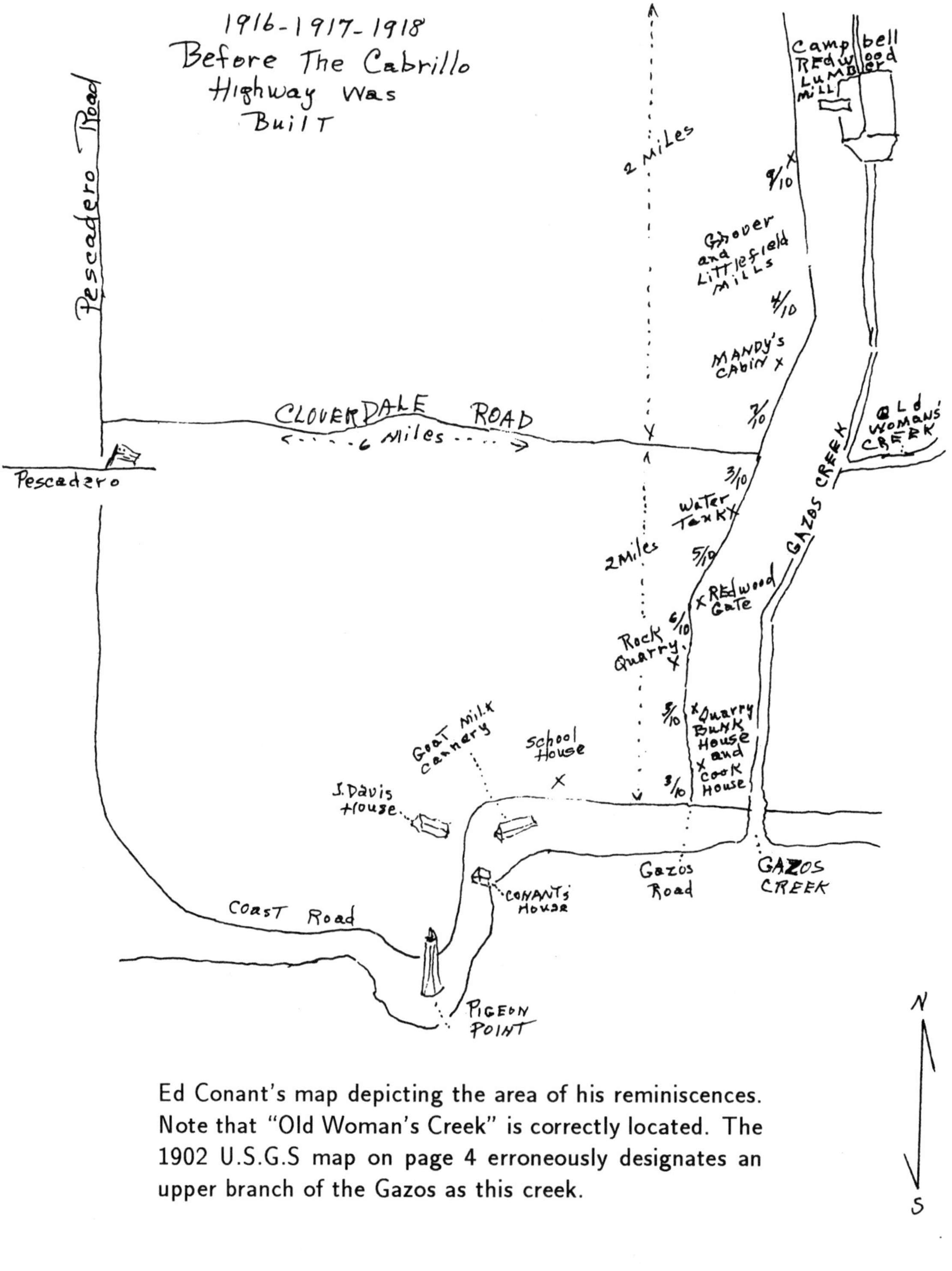

Ed Conant's map depicting the area of his reminiscences. Note that "Old Woman's Creek" is correctly located. The 1902 U.S.G.S map on page 4 erroneously designates an upper branch of the Gazos as this creek.

II

THE GAZOS CREEK ROAD LOCATION

THE GAZOS CREEK ROAD is a road that is well known today. It is located in San Mateo County. It starts about nine miles south of Pescadero at the Cabrillo Highway or Highway One, whichever you want to identify the Cabrillo Highway by today.

When we lived in the Gazos in 1916, 1917 and 1918, it was called the Coast Road which was an appropriate name for the road as it followed the coast. It was a narrow dirt road from Santa Cruz to Half Moon Bay. The Gazos Road intersects the Coast Road a few hundred feet north of Gazos Creek. The Gazos Creek Road runs along the Gazos Creek a little over five miles, then leaves the creek and goes north or northeast up to the top of the mountain which is called Sandy Point.

At Sandy Point there are two other roads. One road is the China Grade Road that was completed to the Gazos Road in 1938 by the C.C.C. the year I lived on China Grade, when I worked for the Charley Johansen Mill on China Grade. The road was closed in later years. The other road is the road that goes down through Big Basin Park. Then the other road goes toward the coast to where it goes over Chalk Mountain or down through White House Canyon right where Dry Gulch heads. The White House Canyon Road goes down through what used to be known by the old-timers as the Low Divide and goes on down to the Cabrillo Highway which is about seven miles plus or minus. Many of these roads have gates across them and you cannot drive through any more.

The Gazos Road used to enter Highway One about one hundred and fifty feet north of where it enters now. At that time, Highway One made a sharp "S" turn where the Gazos road entered Highway One so as to line up with the old wooden bridge across Gazos Creek. It ran in front of the Old Gazos Saloon which burned down in the late 1940s.

When I moved into the Gazos Creek area in 1916, a man named Tony Bennett was running the old saloon, until Prohibition was voted in and he had to close. The saloon was replaced with the Gazos Creek Bar and Restaurant which is there at this writing in 1990. Bob Pinkham ran the saloon for a few years before it burned.

There have been many who have operated the old saloon and the newer restaurant and bar. I don't recall all the names and owners. I remember a few of the names which are: Tony Bennett, Ame Robinson, Bill Arnold, Tony Semas and Bob Pinkham.

There probably was some idle time in between these names that operated the old saloon. I am sure there are a few names that have been forgotten over the years. The restaurant that was built after the old saloon burned down and which is there today, I know has had a few operators that I don't remember.

The mills and the camps were usually many miles to town over a rough road. There was no electricity, no running water, and no refrigerators. It was a life style that I would prefer over today's.

When the mill shut down at night and supper was over, the camp became as quiet as a graveyard. On weekends many went to town and to the old saloons, which today are called bars. So the camp was really quiet on Sundays.

The old saloon was really a headache for the owners of the different saw mills and shingle mills. The mills in those days ran six days a week. Many of the mill men and woods men, as soon as they were paid, would head for the old Gazos Creek saloon on Saturday and come back on Monday if they were not in jail or still drunk and could not work.

I recall one woodsman, a heavy drinker but a real hard worker and a number one woodsman. The boss would go into Santa Cruz every so often, go to the jail, and bail him out. The owners of the mill knew right where to go to find him on Monday morning. I am not saying only the woodsmen and saw mill men were the only heavy drinkers, many of the owners were also.

I have worked with woodsmen that I have seen stay drunk for three and four weeks and never draw a sober breath. These men were hard workers. I have seen it rain three inches in a day. This man would work all day in the rain until night before coming into camp. He worked every day for a couple of months or so, then would go on one of these binges for a week or so then back to work.

The split material camps I have worked in seemed to be worse because some of these camps were fifteen or more miles from town and no transportation and no activity in most of the camps, so they would drink.

The Gazos Road was built with Chinese labor. They used picks and shovels and horses with scrapers to build the road. It was only wide enough for a wagon, with places along it to turn out.

Five miles up the Gazos Road where the Beadle Brothers built their mill in 1917 and where Bear Gulch intersects the Gazos Creek, there was a huge water wheel that was ten or twelve feet high, which was still there when I lived in Gazos in 1916. It could be seen from the Gazos Creek road in those days and it stood about where Bear Gulch and Gazos Creek met. The water wheel was used in the eighteen hundreds by one of the saw mills. I never could find out what it was used for.

It was old and covered with moss. It disappeared in the nineteen-twenties by vandalism. It was probably used for supplying the water into the old water flume that ran down to the Cloverdale Road. They floated lumber down the flume to a landing there. From there the lumber was hauled by wagons to where it was shipped out. Maybe it was used for some other source of power to run a piece of machinery.

I can find no one who ever saw this water wheel except my own family. I saw it many times as it was close to where we had our favorite swimming hole and where I used to fish for trout.

In 1916 the Rock Quarry Camp was located on the right hand side of the Gazos Road. It was just three-tenths of a mile from Highway One. The camp consisted of a cook house and bunk houses, where the quarry men stayed.

This is the same location of the pumping plant that was used to pump water to the Goat Milk Canning Plant Reservoir at Pigeon Point.

The Gazos Rock Quarry was six-tenths of a mile from Highway One up the Gazos, on the north side of the road. You can still see the large hole. Mr. Logan was the black cook for the Gazos Creek Road Gang and Gazos Rock Quarry. He had served time in San Quentin for murder. Dad had known him in the logging camp in the 1880s when he was cooking for the sawmill. He weighed over three hundred pounds and was six feet seven inches tall. I met Mr. Logan in 1916 at the Gazos Creek Quarry Cook House.

In 1916 the quarry had a gravel chute that came out over the Gazos Road to load wagons. It was the only quarry around for miles.

The next roadmark that I remember is about one and a half miles up the Gazos Road from Highway One. Here you come to the first redwood trees. There are three trees growing at the same location today. They grow right along the right side of the road. These trees were known by all the old timers, and they called them "The Redwood Gate," because they were the first redwoods as you entered the Gazos Road.

Next you come to the Cloverdale Road that intersects the Gazos Road two miles from Highway One and runs north to the town of Pescadero. Across from where the Cloverdale enters the Gazos Road, there is a little creek that has always been known as "The Old Woman's Creek." It is a small creek which runs the year round. It was a well known creek when I lived in the Gazos in 1916, 1917 and 1918, for its good fishing. Where the name "Old Woman's Creek" got its name, I do not know, but the same old story is that an old woman lived up there, so they named it "Old Woman's Creek."

The Gazos Creek was easy to get to and it had been closed to fishing for a few years before we moved to the Gazos and the creek was full of trout. It was no trouble to catch fifty trout in two or three hours. We used to fish the Gazos Creek all summer long and never met a single fisherman. There were no cars in those days and the people who lived over there never seemed to be interested in fishing. As the cars increased, that was the beginning of the end of the trout fishing as I knew it.

— Conant Collection.

CAMPBELL LUMBER COMPANY - Gazos Creek- 1917.
Curtis Miracle house overlooking the MILLPOND.

III

REMINISCENCE

THE NOISE AROUND A MILL is really loud, — whistles blowing, trucks, machinery, saws buzzing, all noise we got used to after awhile. But standing at the same mill site today — with only my memories to bring the pictures into my vision — one would never believe such a behive of activity had ever been there.

No smell of smoke made from the burning pit, nor any noise, only the birds in the trees and the breezes swaying and murmuring in the tops of the tall redwoods and the creek still hurrying along to join the salt water of the ocean. The stillness, after a saw mill moves or shuts down, is like a grave yard. I can understand why old abandoned towns are referred to as "ghost towns".

The logs that made the mill pond are still there today, after eighty years or more. Bloom ran the mill and used the mill pond in 1907. Just south of the mill pond you can see the piling along the Gazos Creek. This is where the Campbell Mill sat. There was a fish ladder from there to the top of the dam, so the salmon could go upstream. The dam was somewhere between fifteen and twenty feet high. I never did see a salmon go up that fish ladder. It was steep and not much water in it. The dam had a four foot plug that they pulled out with a cable, when the mill shut down for the winter. The salmon never came up until the winter rains brought the creek up to where they could get up at the mouth of the Gazos. By that time the mill shut down and the pond emptied so the fish could go up anyway.

— Conant Collection.

CAMPBELL LUMBER COMPANY, Gazos Creek - 1917.
Note that there are two smokestacks.

[Facing page.]

CAMPBELL LUMBER COMPANY – Gazos Creek - 1917.
(L to R) Vernon Conant, George French, Lottie Miracle, Mr. Strom, Mrs. Strom, Mr. Duncan. In the backgroung is a Kissel Car. It had carbide lights and the steering wheel was on the right.

— Conant Collection.

I am eighty years old at this writing and I know almost where every cabin was located, also the mill, sawdust pit for burning, blacksmith shop, barn, cookhouse, commissary, skid road, filing shed, water tank, mill pond and bull donkey. There were fourteen of us living at the mill in 1917, all related. There were around fifty or sixty people living at the Gazos Mill. Five of us were going to school from the Gazos Mill.

So, I must write down my memories so that others, who shall read this, can visualize what a wonderful family life I had right around that mill while I was a youth.

I live in Mountain View now and ride over to Pescadero several times a week, then up the Cloverdale road into the Gazos and up to the old mill site. It surely brings back fond memories of living around sawmills and the life I followed a few years later. I never forget where everything was located at the mill, because after we moved I was up there often, fishing and hunting with my father.

As I stand there today, 1990, and think back to 1916, 1917 and 1918, it reminds me of an old song that says "If God would but grant me the power, just to turn back the pages of time". Many of us wish we could do this, because Mother and Dad, my uncles, all my cousins, and my grandmother, are all gone, and only my memories remain.

I have no way of knowing if there are any of them still alive who lived at the Campbell Redwood Mill except my two sisters and brother and myself. I have never met a single person who lived there when I did. — It has been seventy-four years since I moved to Gazos.

[Facing page.]

JIM CONANT and team - Gazos - 1917. Jim Conant's head appears over right horse. Man standing in Gazos road is William Gregory — A saw filer for the saw mill.

— Conant Collection.

IV

LIFE AROUND A SAWMILL

THE LIFE AROUND a sawmill was a hard life, but had its good points too. The mill started at 8:00 A.M. so the horses had to be fed at 6:00 A.M. to have enough time to eat. The fireman at the mill had to fire the boiler and get up steam to operate by 8:00 A.M. There were a number of places in the mill that had to be greased and oiled. So there was a lot of work to be done before the mill started.

There was no extra pay for overtime nor were there any coffee breaks. When the mill started working the place was a beehive of activity, saws a-humming, whistles blowing, and the steam safety popoff valves blowing, along with the sound of the donkey engines running. Today you go there, and there's not a sound, — maybe a bird or a breeze in the trees and the murmuring of the creek running by.

After a hard day's work of usually ten hours or more, you would come in out of the woods, dirty and hungry. If you were married, you would sit down by the wood fire and relax till supper time. If you ate in the cook house they served supper at a certain time, or fairly close to the same time. These bachelor woodsmen would come in, wash up, and wait until the call for supper was heard.

The cook house usually had an old circular saw hung outside and the cook would bang on it. I believe you could hear it for a half mile up and down the canyon.

— Conant Collection.

CAMPBELL REDWOOD LUMBER CO. COMMISSARY, Gazos Creek - 1917. Built in 1905.

The main food was meat, potatoes, beans, bread, coffee, bacon and eggs. Sometimes pies or cookies were served. Some took their hats off and some did not. After supper some went to their cabins and read, some played cards and some had their wine or whiskey.

The commissary was a building about twenty-five feet long. All the staples for the cook house were kept there. The families and the bachelors who did their own cooking could buy from the commissary. In those days you bought by the sack such things as flour, beans, potatoes and sugar. The coffee beans were bought by the sack and ground when you needed the coffee.

Many families had their own coffee grinders, and would grind the coffee beans every morning as needed. Every morning you could hear the coffee beans being ground, as many had the grinders attached to the outside of their cabins.

The commissary bought syrup by the fifty-gallon wooden barrel. We used to buy syrup by the gallon. I remember going down one morning to fill our jug with syrup and it was a cold morning. The syrup was so thick it would hardly run through the wooden spigot. It would take between forty-five minutes to an hour to fill the gallon jug. On a hot day or in the afternoon it ran faster. The syrup in those days was really thick and had a blue cast to it. It was like heavy molasses.

This was just the way of life in the early nineteen hundreds. We did not mind it as we never knew any difference. The closest town was Pescadero, and that was sixteen miles over a rough dirt road.

The commissary got its supplies when my dad would take a team of horses and wagon and go to Pescadero for them. That was a sixteen mile round trip. It took most of the day.

I recall, in the early spring of 1917, when the Campbell Mill was getting ready to start operating, Mr. Bloom, his brothers and his men came in with horses and wagons and started taking out the machinery from the mill, which he claimed was his.

They loaded some pieces of the mill on the wagons. I remember them tearing the lumber rollers out of the mill and loading them on the wagons. When Mr. Bloom got up into the wagon to start, one of the Campbell Mill men stepped in front of the horses and held them so they could not start. It made Mr. Bloom so mad he got down from the wagon ready to fight, and when he did, the man stepped out from in front of the horses and they ran away scattering machinery in the road. One horse fell down and the wagon and horses almost went into the mill pond. It was like a scene out of the old western movies.

— Conant Collection.

JAMES CONANT — Gazos, 1917. Horses - Tom and Jerry.

While Mr. Bloom and his men were taking the machinery out of the mill, the Campbell men went down the Gazos Road and felled a few trees across the road to block the Bloom men from getting out before darkness set in.

I remember watching one group felling a tree and they were in such a hurry it got away from them and fell up the hill, instead of across the road. They finally got enough men to stop Bloom and his men before some got injured.

Things finally got settled and the Campbell Redwood Mill started operating in the spring of 1917 and part of 1918.

In 1916, the Gazos Road went up Gazos Creek a little over a mile past the Campbell Mill. The Campbell Mill was being worked over and also the skid road out to where they were to start logging. This skid road was quite long. It followed right along the Gazos Creek on the south side. In many places along the creek, they had to use logs to shore up the skid road.

You can still see a small section of the old skid road and a few logs which they had used for shoring at the time of this writing — 1990. Many of the logs used in the skid road were in shape and had to be replaced. The logs used in the skid road to skid the logs over were around ten inches in diameter and sunk in the ground, so as to leave about one-third of the log above the ground. These logs were placed about four to five feet apart. This was a big job, as it was all done with picks and shovels. Once in a while they had to replace a skid road log when it got damaged. But on the average run these logs lasted quite some time.

[Overleaf.]

JIM CONANT'S hunting dog QUEEN - 1916. Campbell Redwood Saw Mill. Note the dragsaw at the edge of the mill pond.

— Conant Collection.

V

THE BULL DONKEY AND THE WHISTLE PUNK

THE BULL DONKEY THAT PULLED the logs over this skid road was stationed at the mill. It was called a bull donkey, because it was larger and stronger than the logging donkey in the woods. The number of logs it would pull depended on the size of the logs. The logging donkey was a smaller donkey, as it would pull one or two logs off the mountain, down to the main road. This was done by block and tackle and consumed lots of time and hard labor. The logging donkey was small and could be moved fairly often to different locations.

When they got enough logs lined up for a haul to the mill they had what they call "chockers," which attached one log to another. Around six to eight logs was the average load. From the bull donkey, stationed at the mill, out to where these logs were chockered together to be brought over the skid road to the mill was a long distance with many turns. You could not see when the crew in the woods was ready to start pulling. They had to install two telephone wires out in the woods. At the bull donkey at the mill, they had a bell operated by batteries.

When the crew was ready, they had what was called a "whistle punk." He would touch two wires together which rang the bell at the bull donkey. Two rings was to start pulling. One ring was to stop. Then the bull donkey operator at the mill, after receiving the "go ahead" signal, would blow the whistle so the men in the woods could get clear of the logs and cable. Men have been killed by not getting clear of the logs or giving the signal before they could get out of the way.

I remember a logger that Mother gave a ride from Highway One up to the mill. He rode with us kids one afternoon going home from school. The next day in the afternoon he was killed. They gave the "clear" signal before he could get out of the way and he was crushed against a stump.

A donkey engine took two men, an operator and a fireman. The fireman kept up the steam pressure and helped repair the breakdowns. The fireman on the donkey engine had other work to do. He had to split wood to go into the donkey engine firebox. There was a lot of greasing and oiling to be done. He also, with the donkey operator, watched the steam pressure in the boiler, as well as the water level. These logging donkey engines, operated with a high steam pressure, dropped fairly fast, so they had to wait until they built up steam pressure to make another pull.

Everything in those days worked slowly compared to logging today with Caterpillars. Moving and setting up block and tackle took time. Getting water for the boiler, for the donkey, was not always handy. Then, sometimes when they were pulling a log off the mountain, it would get hung up between stumps. Sometimes the donkey operator would try to free a hung up log and would snap a cable.

This was time consuming and expensive and the margin of making a dollar and going broke was very small. Most of the sawmill owners didn't have too much capital to operate on. The Campbell men went to San Francisco and bought old cable that the cable cars had discarded. The old cable kept breaking and many man hours were lost waiting for it to be spliced. My dad had that job. The mill could keep operating because they had logs in the mill pond but the logging was delayed until the cable was repaired.

[Facing page.]

Stationery Bull Donkey engine. Gazos Creek. The man with the two five-gallon cans and yoke was water carrier for the donkey engine.

— Harvey Mowry Collection.

New steel cable was expensive and the differential in the price of new and used cable amounted to hundreds of dollars. In those days, the difference in a few hundred dollars was a matter of staying in business or going broke.

When they wanted to move the donkey engine from different settings to get to where the logs were, they had to hook the cable around a large stump or a tree or use block and tackle. A donkey engine can pull itself up a very steep grade. This was very dangerous, as a cable could break and lose the donkey.

In the woods when they have all these logs on the main skid road ready to be pulled to the mill pond, the stationary bull donkey located at the mill pulls the logs over the skid road to the pond. The reason they had a mill pond at the Campbell Mill was because there was no place to store the logs as the mill was right in the bottom of the canyon. So a mill pond made a good place to store the logs until they could be sawed. A mill pond had another purpose. It washed most of the dirt off the logs so it was easier on the saws.

When these logs were pulled in from the woods over the skid road by the bull donkey they had some logs set at an angle to veer these logs off the skid road into the mill pond. Occasionally they would get a log that would not float. It was the butt end log, or the first fifteen or twenty feet of the tree or more, the heaviest part of the tree. The logs that did not float but went to the bottom of the mill pond were called the "sinkers." A winch was set up on two logs and a cable that they worked around the log raised it.

The bottom of the tree, which is the heaviest part, always makes the best lumber. It will not rot as fast when put in the ground as lumber made from the upper part of the tree.

The mill pond man was always busy as he had to ride the logs from the center of the pond and bring them over to where he had a dragsaw mounted on a platform at the edge of the pond. Here he sawed the logs into different lengths. It depended on the order.

I saw this man fall off the logs many times into the pond. It was not easy for him to swim with his clothes and logging boots on. The mill pond man's job was to float these logs over to the car that ran from the mill down a track into the pond. This car had spikes, so when the log was floated over it, the spikes held the log from rolling off, as the car was pulled by cable up out of the water into the mill.

Then a cable was put around the log on the car and it was rolled off onto the carriage. Here the log was sawed into lumber. All the work to get the logs from the woods to the mill pond and then into the mill and onto the log carriage was very slow and hard work. Labor was cheap, but so was the sawed lumber.

The lumber went down on rollers, where it was cut to different sizes. From there, it was put on a tramway car and pushed down the track where the car was unloaded by hand.

The lumber was graded as they stacked it. From there two men would take it from the stacks of lumber and stack it on two large sawhorses. Then a lumber truck would back under the load. From there it was lowered on a truck by two large jacks. The trucks in those days had solid rubber tires and were not very large, so the load of lumber was not too large either.

— San Mateo County Historical Museum.

Timber fallers near La Honda. The undercut is finished and they are ready to start sawing the backcut.

VI

THE METHOD OF FALLING TIMBER IN THE 1800s

THE METHOD of felling timber in the 1800s remained the same until about 1940 when the chain saws became popular. I began working in the woods before the days of the chain saws. So there was no change in the tools used for felling timber for better than a hundred years.

The only major change was from the two-man crosscut saws to the chain saw. I speak only of the way I worked, as I worked mostly in redwoods and some fir. There is a vast difference in falling redwood trees, than falling fir trees. The tools are the same for falling redwood as for fir. The difference is in the setting of the teeth of the falling saw.

When I used to sharpen my crosscut saw, I used to bring it out of the woods up to camp. Every camp or mill had a saw "filing bench". Some woodsmen made their own filing benches in the woods, as in some camps it was a long way to the woods. It only takes a couple of posts and three or four boards to make a filing bench.

If you were working in the redwoods you had to make sure when filing your crosscut saw that you had the rakers on your saw set lower than when you were working on a pine log or fir. If the rakers are too high your saw will pull very hard and will not remove the sawdust from the log correctly and the saw will not cut right. Now on a fir log you have the rakers set higher because the wood is harder and the saw does not cut as fast. Then you would have the sawdust accumulating in the cut and not being removed by the rakers.

So your saw needs to be sharp and the rakers filed to the right length and the teeth set to the right clearance. Even at that, it is slow, hard work cutting through a five or six foot log. We have what is called a "raker gauge" for setting the rakers, and a "saw set" or "spider" for setting the cutting teeth clearance.

The reason you have to be more careful in falling a redwood tree is, you have to select the best ground on which to fall the tree. By the best ground, I mean no knolls or ravines. A redwood tree is soft, and if the ground is not right, you could very easily shatter the tree. If you do, the tree is practically lost. Sometimes there is no decent ground. If possible, you can fall a smaller tree in the low place to make a better bed for the tree to fall on.

If the redwood tree is four or five feet in diameter, it will shatter very easily because a larger tree, as a rule, is of a finer texture and of a softer wood. A redwood tree that can be felled for lumber today is as small as eighteen inches in diameter. A redwood tree of eighteen inches in diameter has a wide grain that makes poor lumber. These small redwood trees can be felled most anywhere and they spring like a rubber band without much damage. I have fallen smaller trees just to make a bed for the larger to fall on to keep it from breaking and shattering.

When I fell a four or five foot redwood tree, I used to "ring" the tree, and that is chopping the bark away around the tree where you are going to saw. Not all fallers ringed their trees, but it made it a little easier for the saw. Bark has a tendency to dull the saw to a certain degree. With the chain saws, they cut through the bark and all.

On the larger redwood trees, some have a tendency to swell out at the bottom and the wood is burley and tough. On trees like this when you are using a cross cut saw for falling, before the days of the chain saw, it was always easier to go up higher on the tree to where the tree starts to swell out or get larger.

To get above this you would save cutting through this burley wood and save a lot of hard sawing. With a chain saw, that is a different situation.

I had some logging done for me in 1975. They never even had a spring board or falling wedges.

To get above this swell in the bottom of the redwood tree we had what is known as spring boards. I am sure you have seen those notches or holes in the redwood stumps where these spring boards were placed or set. Sometimes you would be up on the side of the tree anywhere from a couple of feet to fifteen feet. That depended on the tree or the steepness of the mountain you were working on. A spring board is nothing more than a piece of wood six to eight inches wide by four feet long with a metal tip bolted on that was round with a small lip curled up to go into the notch on the side of the tree.

Sometimes you would place a number of spring boards around the tree and place a plank from one board to the other to make a platform so you could walk around the tree. I used to make mine out of the sap from a redwood, and the blacksmith made the iron that bolted on the end of this five foot sap board. The board was about five or six inches wide.

We used to chop a hole in the tree in which to set the spring boards. The iron tip on the end of the board was in the shape of a horse shoe, with two straps of iron about three inches wide running back to bolt them onto the spring board.

We used a felling axe to chop the holes in the tree. A felling axe has a narrow double blade. We chopped the holes about three to four inches deep and on a downward slope, so when the spring boards were put in the holes, the other end of the spring board was higher than the hole in the tree.

The reason for this is, when you get on the spring board, the lip and weight should bring your spring board down to about level. Otherwise, it would slope downhill after you got on the board and that would be very uncomfortable to work.

Sometimes we had to set a number of spring boards in one tree to make a platform to walk around the tree. If the tree is on steep ground on the downhill side, you could be up in the air ten feet, sawing in and setting your felling wedges.

On the other side of the tree you may only be up two or three feet from the ground to where you are putting your under cut, that is if you are felling the tree up a hill.

The direction you are felling the tree depends on the ground and how much the tree leans. So the circumstances have a lot to do with the direction the tree is to fall. You are limited to such, as if a tree leans heavily downhill and you have better ground for the tree up the hill, you will have to let it fall in the direction it leans, and hope for the best.

I will make the comparison with a tree that has no lean, with the leaner. To fall the tree that has no lean, you set the spring boards, if needed. Then I used to ring the tree by cutting the bark away to where I am going to saw into the tree for the undercut. A tree that has no lean to it is sawed first on the side where you are going to put your undercut or on the side of the direction the tree is to fall.

We will say we are falling a tree four feet in diameter. I always like to saw into the center of the tree. If the tree has to fall exactly where you want it, then you should saw into the width of the tree. Next, you are ready to chop the undercut. You chop the saw cut until you come to where you stopped sawing. This should be cut fairly clean. Now, you have the undercut in and I should say this chopping out for the undercut is done with the falling axe. A falling axe is a double bitted axe and made thinner than a chopping axe. The handle is longer than an ordinary axe, so as to reach out farther in cutting the undercut and into the center of the undercut.

This axe was used only for falling, and it is kept really sharp. A falling axe is thin and kept sharp because redwood bark is tough and springy. An ordinary axe just bounces off the bark.

So now we have the undercut in; the next step is to go to the opposite side of the tree and about twelve inches above where you sawed for the undercut. You start sawing. After you have sawed in far enough, you can set your falling wedges in so the tree won't settle back on the falling saw. You saw a little, then you keep taking up the slack with your wedges.

If the tree leans a little toward the undercut, you can fall the tree without any wedging. Again, there are things to consider, such as is it critical where the tree falls and is there wind to break the tree back across the stump. All this knowledge only comes by experience, not by books. I always felt like I had better control of the tree by setting my wedges. You can saw into about three and one half inches of the undercut, then you can drive your wedges and the tree will fall. This cut, where your wedges are set, should be square or equal on both sides.

When you saw up to three and one half inches or four inches of the undercut, that is called by a faller as "holding" so much wood. Now, if you wanted to pull that tree a little to the right for some reason when you are sawing in where your falling wedges are, you would saw into about two and one half inches on the left of the undercut and about four inches on the right. This would be holding wood on the right side of the undercut. When the tree starts to fall, the side that you are holding more wood on would pull a little in that direction. These numbers of inches are only approximate. Other things intervene and would change to more or less inches.

"Gunning sticks" are two sticks ten feet long, hinged at one end. Place the other ends, one on each side of the undercut. You stand in the center of it. The tree will fall in the direction to which the hinged ends point.

The old timers used to place the felling axe in the center of the undercut. That would give them the direction, but it was not as accurate as the gunning sticks.

Now, in falling the tree that leans down hill, and you are going to fall it up the hill, you would reverse the procedure from the tree that has no lean.

With the leaner, you would first saw in the opposite side of the tree where you are going to put the undercut. If you were to put your undercut first, you could not saw in deep enough to set your wedges. It will bind the saw because the tree has been weakened. If you saw in first, before you put in your undercut, you can saw in deep enough to where you can set your wedges and shims.

After you have set the wedges on the opposite side of the tree deep enough so you have room for the falling saw, then you go to the opposite side and make your undercut. These falling wedges are about seventeen inches long and tapered from a sharp edge to one inch in thickness. If the wedges were any thicker they might bounce out when you start to drive them, and you are picking up tons of weight. In wedging a leaner that is around ninety feet to one hundred feet tall if you raise the tree an inch at the undercut, you have moved the top many inches.

You never use one felling wedge in raising a leaning tree, but three or four. It depends. As you saw in, you have steel "shims" that the wedges are placed between, so the wedges won't sink into the wood while you are driving them. These steel shims are different thicknesses, but very thin to start with.

When you get the tree raised, say four inches, you have to be very careful and follow up with other wedges as you drive the wedges in. Sometimes you drive on a wedge and it will bounce out. If you do not follow up with another wedge or wedges, the tree could come back with all that weight and break across the stump, and the tree would be lost.

I have had redwood trees raise eight inches or more and still not fall. It takes hours sometimes to raise a tree that is a leaner.

The woodsman in my day had names for redwood trees. When they came to a good sized redwood that was all burned out in the center and maybe only four or five inch rim around the outside that held the tree and furnished enough nourishment to keep the tree alive; that type of tree was called a

— Conant Collection.

EDWARD and VERNON CONANT making shingle bolts – notice the dragsaw. CECIL GIBSON'S SHINGLE MILL near the Saratoga Summit and Oil Creek - 1946.

"goose pin." Then there was the redwood tree from which the top had been broken and the limbs grew out around the top where it broke off. So it looked like a big umbrella. So they called the tree an "umbrella tree".

The redwoods that they are falling today are eighteen inches in diameter and they were called large "suckers" in early days. I don't know if such trees have special names today, as it has been over forty years since I have worked in the woods or mills.

VII

GAZOS CREEK SCHOOL

WHEN I WAS GOING to the old Gazos Creek School in 1916-1917-1918, the Gazos Creek Road came out on Highway One a hundred feet or more north of where it enters today. The old Gazos School I went to looks like the one built in the 1800s except for the small portion added on the front with what was called an "anteroom". That is what it was called in those days. I cannot find anyone who knows, but the pictures I have of the new school looks like it was added on to the old school.

Our drinking water in those days was packed in buckets from the Gazos Creek and put in a crock-like dispenser with paper cups to drink from. I don't remember how they decided who was to go after the water. I know I packed a few pails of water.

I remember the teacher coming to school in the old horse-drawn cart from down near Año Nuevo with one of the Steeles where the teachers stayed. I don't remember if it was at Stanley Steele or Bernice Steele's home. I also remember Mildred Steele and Ella Steele. I remember one morning when they were coming to school, the horse either balked or something happened, I am not sure, but Miss Powell, our teacher, got her arm broken, so we had no school. I don't believe Miss Powell taught there after that. I don't remember.

[Facing page.]

Gazos Creek School about 1885. Adrian Manley and Grover Steele among the children on the step.

— Lottie Manley Green's Collection.

— Norman Steele's Collection.

Miss Jackson — teacher at the Gazos School — 1897.

[Facing page.]

The old Gazos Creek School — located on the property where the Campbell Mushroom Plant is now (1990). When it burned down, the new school was built at the mouth of the Gazos Creek. Norman Steele appears second from the left, Rensselaer Steele Junior, third from the Left. Edward Conant and brother and sisters attended the old school in 1917.

— Norman Steele's Collection.

I remember the names of the four teachers I had while I was there. They were Miss Powell, Miss Bull, Miss Bullwinkle, and Miss Gleason. I believe they were all single. I am not sure of that. I believe there were twenty some odd children going to the school when I was there. I will endeavor to give you a few of their names. The children who came from up the Gazos Creek and the Campbell Saw Mill in the Gazos were: Pearl Conant, Edna Conant, Vernon Conant and Edward Conant. Ida Maurica, Maggie Maurica and Harry Newcomb were there. Then Silvas who lived on Highway One just north of the school were also there. I stand corrected here, but I think their names were Carry and Tony; then there was a Silva up near Pigeon Point. Then from Pigeon Point there were John Nixon, Isabel Davis and her two sisters and Alden Robinson, then the Vegas from over on Bean Hollow. There were two or three of them. One of the girls' names was Julia. I don't remember the others.

Then there were the Pinkhams: Elbert Pinkham, Lester Pinkham, Robley Pinkham, known as Bob, and one girl Lila Pinkham. There may be more and I am sure there was, but over the span of seventy-four years, I think I am doing good to remember these.

The old school that I went to was on the hill along Highway One. It burned down, I believe in the 1920s, then they built the school in the Gazos area which still stands at the time of this writing, but was converted into a house by Bob Pinkham which now, at this writing is owned by the Campbell Soup Company.

I remember in 1916 when Mother would take us to school with the horse and buggy. The ruts in the road were so deep one stayed in the ruts. The mud was so bad in these ruts that the wagon wheels would pick up big chunks and it would drop off the wheels.

Many a day I walked four miles home from school which added up to eight miles a day. There were no automobiles to speak of, as I never got a ride in a car in all those three years, only a truck and then not very often. There were trucks for hauling, but there was only one truck that I could ride, as the

— Conant Collection.

CAMPBELL REDWOOD LUMBER COMPANY, Gazos – 1917.
(L to R) Vernon, Pearl, Edward and Edna Conant. In the background is their cabin built in 1905. Wooden water flume runs across the center of the picture, carrying water from the spring to the water tank. The water was for the camp and also the mill.

truck drivers had only one jump seat. The rest of the truck was just a frame and no place to ride safely. But any ride looked good when thinking of that four mile walk ahead after school.

Our old horse got used to listening for the trucks. He kept putting his ears back and forth listening for them. You could hear noise in those days a long way up and down the Gazos. You could hear their exhaust whistles, as there was no noise from airplanes, nor cars.

— Photo by Lottie Miracle.

Edward Conant, Age 7 — Gazos Creek School — 1917.

In the winter, it was so cold the horses' watering troughs were frozen over. Icicles were hanging from the branches of the trees.

The trucks that were hauling lumber were the Autocar, White, Liberty, Moreland, Indiana, and Diamond-T. They could only haul in the dry weather. No one can visualize how the Gazos Road was in those days unless you lived there, as I did in 1916, 1917, and 1918. In the summer it was rough and dusty, as the lumber trucks hauled lumber out from the Campbell Saw Mill and the Beadle Mill.

The lumber was hauled to Pigeon Point Lighthouse, where it was unloaded in stacks beside a small railroad track that ran out to the wharf. From there, they took the lumber piece by piece and stacked it on the little flat cars that were pulled by an engine, made from a Metz automobile, out on the wharf to be loaded on the Beadle boat, called "The Cyclone," and then shipped to San Francisco.

Mr. Beadle, who owned a saw mill, one mile above the Campbell Mill in the Gazos, also owned a steamship company in San Francisco. He brought his boat down the coast to Pigeon Point and anchored a hundred yards or more out from the wharf that was there in those days.

Now in order to get the lumber from the wharf out to the boat, they had to get a cable to the boat. They had a small cannon they loaded with black powder and a steel ball, with a rope tied to the ball. They would shoot this ball over the boat. On the other end of the rope they hooked a cable and pulled it out to the boat. After they got their cable and rigging hooked up, they pulled the cars loaded with lumber out to the wharf where they put a sling around the lumber. It would go down the cable to the boat, where it was let down into the boat. It took a number of days to load this small lumber boat. It was a very slow operation.

The boat was lower than the wharf so the lumber and sling traveled to it over the cable. Rigging hooked to the sling brought the sling back. They had a donkey engine at the wharf to let the sling and lumber down to the boat slowly and bring the sling back.

VIII

I LIVED AT PIGEON POINT IN 1918

WE MOVED OUT of the Gazos up to Pigeon Point in 1918. From there the Beadle Boat took the lumber to San Francisco and stacked the lumber on the wharf. From there they had to load it onto trucks and wagons by hand. Then they took the lumber to the lumber yard where it was restacked and then to the customer. So you see what a procedure it was to get the lumber from the tree to the customer and the expense of so much handling. There was no profit left.

One day they were loading the boat when a huge storm came in. They only had the boat partially loaded. They had to cut the anchor ropes and get out to sea, which was very dangerous. If the ropes got tangled in the boat's propeller they could end up on the rocks. The breakers and the wind were so treacherous you could look clear under the boat and see the propeller.

[Facing page.]

Loading Beadle's boat at Pigeon Point.

— Fanny Steele Collection.

— Ann Steel's Collection.

Keepers house — Pigeon Point 1872.

[Facing page.]

Aerial view of Pigeon Point — 1920s.

— Ann Steele's Collection.

— Conant's Collection.

Beadle's boat named "CLEONE" at Pigeon Point Wharf — 1917. Note the row boat on the beach. Edward and Vernon Conant played on this beach in 1918.

[Facing page.]

Pigeon Point — 1914s.

— Conant's Collection.

— Ann Steel's Collection.

Pigeon Point Wharf — early 1900s. Grain storage shed, lower right.

[Facing page.]

Pigeon Point Grain Chute. Grain was loaded into boats for transport to market.

— Fanny Steele's Collection.

I was living at Pigeon Point when World War One ended and the terrible flu was raging. We wore home made masks from many thicknesses of cheese cloth to prevent us from catching the flu. I don't think it did much good. We were told to burn sulphur on top of the old wood stove and breathe the smoke fumes and that would kill the flu germs. We tried anything and everything as there was no medicine in those days that you could get or take for the flu epidemic. Many died from the flu. I lost two cousins from it. Our family of six all had it but one. We all made it through.

I remember when we received the news that the war was over. I watched one of the lighthouse keepers climb to the top of the lighthouse where the little round dome is and anchor the American flag to it.

Some time in 1918, shortly after the war was over, there was a motorcycle race from San Francisco, down the El Camino Highway to either Watsonville or over the Santa Cruz Highway. From Santa Cruz, then up Highway One or Coast Road, as it was called, to Pescadero. Then on to San Gregorio from there. I am not sure if they went to Half Moon Bay or over the La Honda Grade, then back to San Francisco.

These motorcycles had side cars and two men. They made three round trips, I think. The one who made it the fastest won the race. This was all dirt road and they went as fast as they could. There were no speed limits and they used the road for a race track. Can you imagine using the main road for a race track today? One would not see a single car all day long on the Coast Road.

[Facing page.]

Mary, daughter of Joe Davis. — Beadle Boat in the background. – 1918.

— Fanny Steele's Collection.

We walked to school, from Pigeon Point to the Gazos Creek School. All week we traveled both ways and never met a single car. So there was not much danger using the Coast Road for a race track. I am still trying to find someone who remembers that Motorcycle Race of 1918, but have had no success.

From Pigeon Point we moved about a mile north on the right of Highway One up on the hill across from where the Pheasant Hunting Club was. In 1919 a lot of that land was planted into sweet peas for seed for Ferry Morse. Also grain and garden peas were abundant along with pink beans. Then we moved to Manteca, and back to Mountain View in 1920.

In the early 1930s I went back into the saw mill work, also shingle mills. I felled timber and worked at cutting split material, such as I stated before, posts, pickets, grape stakes, ties and shingle bolts. I also did some shingle packing, gulching shingle bolts and splitting out shingle bolts and a little pack mule work. All in all, I have some fond memories of the past, also of the hard work that young people will never experience today.

But still they were "the good old days".

IX

HOW I LEARNED TO BE A WOODSMAN

I HAD A GOOD woodsman from whom I learned the trade. My father was a number one feller and logger.

My father, Jim Conant, started working around logging camps in 1888, as a boy of fourteen. He started as a skid road greaser for the bull teams. It does not sound like a very important job but it was, for you had to know when to add grease to the skids so the bull team could pull the logs and when not to, and how much to grease. By adding too much grease on the down grade the logs could break free and slide into the bull team injuring or killing them.

My dad worked as a skid greaser for a few years and then he became a bull team driver for different mills. Dad also worked as a logger and timber feller. He was an experienced, all around saw mill worker.

When I started following the woodsman life in the early 1930s, I met a mountaineer whose name was Pat Coyne, who was known from Santa Cruz to Half Moon Bay. He worked all through the Santa Cruz Mountains. Mr. Coyne was well known around Pescadero. He was a number one woodsman at felling timber, making posts, pickets, grapestakes, ties and shingle bolts. I worked with him up on China Grade for Charley Johansen.

I lived next door to him, so I had a good instructor. Then I worked with a Swedish salmon fisherman from Alaska. He was a really hard worker and a good woodsman. I learned a lot from him. His name was John Olson. The only thing I did not learn from him was, when it poured down rain all day, to keep

— Soquel Pioneer & Historical Society. (Gift from Jim Conant's Collection.)

Jim Conant driving "Bull Team" at Grover's Mill near Soquel.

on working. John never stopped for the rain. The only thing that would stop him was after he worked a couple of months he would go on a two-week drunk, eating nothing, just drinking. He eventually got cirrhosis of the liver and passed away in the 1940s.

It was before the days of the chain saw when I worked in the saw mills falling fir trees and redwood trees. The size of the trees were much larger than the trees they are falling today for lumber. We had many trees five, six, seven and eight feet in diameter. In my day, in the woods, they would call an eighteen inch diameter tree a "sucker." A redwood tree this small makes very poor lumber, because by the time the bark is removed very little redwood is left. Only sap and a very wide grain piece of lumber remains. If the sap is put in the ground it will decay very soon. The trees are so small in diameter that a lot of sap is left. It is important today to save as much as possible because the price of stumpage is very high.

When I was working in the mills a four or five foot redwood had a finer grain, because of its larger size. A tree that size made number one lumber, because they removed all the sap and the larger the tree, the finer the grain and the softer the fiber as a rule, not always.

Redwood sold for five to seven dollars a thousand board feet. That was for standing timber which is called "stumpage". A man who is called a "timber cruiser" would go through the woods and estimate how many thousand or million board feet of timber. Now days, this is done at the mill as the logs are hauled in.

Working in the shingle mills or saw mills, you used the same tools for falling. For the shingle mill you worked your tree up in fifty-one or fifty-two inch lengths, and then would split the log into what is called a shingle bolt. A shingle bolt will make one bunch of shingles. In order to get a bunch of shingles from one shingle bolt, the bolt had to measure at least twelve inches by fourteen inches. At the mill, the end of the bolt had to be squared on the boulder saw, which took a couple of inches for squaring. A shingle length is sixteen inches, so a bolt had to be forty eight inches. That made three shingle blocks out of each bolt which made a bunch of shingles.

The same goes for shingles, the larger the tree, the better the shingle material because a larger tree has a finer grain and a better chance of the wood being softer. A two foot diameter tree which today is made into lumber would not make good shingles. It would have too large a grain and would stand a better chance of the wood being hard. A tree at least three and a half feet in diameter or larger is the desirable size.

When cutting shingle bolts or making posts, pickets, or grapestakes, one had four or five splitting wedges. They were about eight or nine inches in length and tapered from about one inch to practically nothing. The reason for such a small taper is, if the taper is too radical when the wedge is driven into some trees they have a tendency to fly out, especially if the wood is a little hard.

In making shingle bolts the length of the cut would be about fifty-two inches; posts, eighty -four inches; grapestakes, seventy-two inches. Posts, pickets, and grapestakes may vary a little in length according to order. Shingle bolts never change in length.

Pickets were two inches by three inches in size, and six or seven feet long. Grapestakes were two inches by two inches by six feet. Posts were four inches by five inches by seven feet. They would vary in size, because on any split material some trees had a tendency to run in and out in seven feet, depending on the tree. In the 1930s I received a cent each for grapestakes. Three cents each was paid for pickets. Seven cents was paid for posts, ten cents for shingle bolts.

The advantage the shake maker had over the one who made posts, pickets, grapestakes, shingle bolts, or the tie maker, was it took a lot more time to work a tree into shakes. He made more money per tree. He was not taking up his time in felling a tree which is time consuming. In those days before there were chain saws, felling was all done by hand, with a cross cut saw.

The last redwood tree that I fell was eight feet in diameter. It took three days to work it up into shingle bolts. A tree that size, for a shake maker, took weeks to work up, but his profit was much greater.

After you cut your log to the length for what you are making, you have split the log in half. A redwood log has what is known as a "wind shake." A wind shake is nothing more than a natural split that runs from the center of the tree toward the outside. These wind shakes vary in length. One should always use this wind shake to split the log in half. Where the wind shake ends you use what is called a "checker." You take the checker and start where the wind shake ends and drive the checker in about an inch or more. Then knock it out with the sledge hammer and do the same over again until you get a deep line from the wind shake to the outside of the tree.

Then you set your wedges in the line made with the checker. In a log four feet in diameter I used, as a rule, four wedges to split the log in half. Sometimes the number of wedges may vary according to the tree and the size of the wind shake.

Some trees will split in a fairly straight line where the checker mark is made. Then again, you will find a tree that will not follow the checker mark very well.

The wedges I used were about four inches wide and tapered from one sixteenth to around one inch at the head, six to eight inches long. The wedges were called splitting wedges. They were different from the falling wedges. The falling wedges were around eight inches in length, and tapered from one inch to a sharp edge at the other end.

When all the wedges have been driven in and the log does not split in half a glut can be used. A "glut" has the same shape as a wedge, but much longer and thicker. These gluts are usually made out of Valparaiso Oak, which is a very tough oak. It grows in the Santa Cruz Mountains.

In making the glut you make it like a wedge, and cut a neck at the top of it about three inches long. Then you place a metal collar around the neck of the glut. This type of oak is so tough I had a glut made of it and I used it a number of years in the woods, and it is still in good condition.

A "crow bar" is then used to pry the log apart. Then wedges are set along the sap line of the redwood log to remove the sap and bark. If one is making posts, the logs are quartered. Afterward at least four posts or more are split at a time if possible. Then you split these into a single post. If you try splitting one post at a time off a log it will taper too much. Redwood trees vary. Some split really well, others are tough and stringy, and some in between.

X

THE TIE MAKERS

IN the 1880s, up until the 1920s, there was a big demand for railroad ties and they called these men, the "tie makers," as well as tie camps where they worked. Spreckles Sugar Plant in Salinas, as well as Hawaii, used to use the ties in their spur tracks. Redwood ties were used as they lasted a long time in the ground, but the spikes that held the rail to the ties had a tendency to loosen from the movement of the loads because the wood was soft.

The pine or fir ties hold the spikes so much better as the wood is harder. So the redwood ties and hand split ties are a thing of the past in the Santa Cruz Mountains. The ties now are fir and are sawed. In making redwood ties we had to split them out by hand. Then we had what is called a "broad axe" for squaring them. A broad axe is around twelve to fourteen inches wide. It is an axe, but quite different from a regular axe.

XI

THE SHAKE MAKERS

THE OLD SHAKE MAKER used to look at a redwood tree and he could pretty well tell by looking at the bark if it would split well enough to make shakes. If the bark was straight with long ribs without the bark crisscrossing, it probably would split well. It is not a sure sign, as I have worked up trees that were just the opposite. The best way the old shake makers and split material workers and I used was to cut a notch in the tree up from the ground about four or five feet high and remove a three or five inch piece of wood. Use a pocket knife and split this sample and see if it is stringy or not.

Shake makers were different from the men who made split materials. They were different in the sense that a shake maker would, as a rule, do nothing else but make shakes. That was his trade. He would go through the timber first and mark trees he thought would make shakes. He would take samples of the trees and he could tell if the tree was a shake tree or not. A tree may be a real good splitter, but it will not make shakes. It takes a special tree for shakes. So the shake maker had the pick of the timber first.

The virgin timber was where the shake trees were found. Since they are cutting second and third growth redwood trees now, the split shakes are a thing of the past. The sawed shakes are used now.

The tool used to split the shakes was called a "fro." It had a long blade about fourteen inches long and a short handle about twelve inches long. The blade was blunt on one side where they hit it with a wooden mallet. The other side was very sharp. When the fro was driven into the block of wood, the handle was pulled sideways and this was the way shakes were split.

XII

TAN BARK

A MAN BY THE NAME of Dresti had a crew cutting tan bark in the spring of 1917 in the Gazos. They were cutting about a half mile before you get to the Campbell Mill on the south side of Gazos Creek. They had many cords of tan bark stacked along the side of Gazos Road. They loaded it on trucks and hauled it to Swanton or Santa Cruz. There were three large tanneries in those days. One was in Redwood City, one in Santa Cruz, and the other one in Santa Clara. The tan bark was used for tanning leather. It was peeled from the Tan Oaks growing in that area. Today they use an entirely different system of treating leather.

I remember the mule pack train that hauled the tan bark out of the woods. They had what is called a "pack saddle" made out of iron for the mules and they loaded the bark on both sides of the mule. They had to make trails out into the woods for these pack mules. When they loaded a mule they had no way of telling how much weight they had on him. So the way they gauged the weight was to load a mule until its fetlocks would touch the ground.

Tan bark was very heavy and the amount was enormous that they loaded on each mule. Pack mules are peculiar animals. Some of them if loaded too heavy or for some other reason, would not move. The pack train driver would get next to the bank and if the mule refused to move, he would push the mule off the trail with his load. This is the way the mules were broken from being rebellious. Mules are a lot smarter than people give them credit for. If things are not just right they won't move. Sometimes they moved when you did not want them to.

I remember when I first started packing out posts, pickets, and grapestakes with a mule for Cecil Gibson on Oil Creek. The mule knew I was not experienced and every chance he got he would run away. If I loaded him too heavy he would run away. He always ran back to his barn. After I realized what was wrong, I never had any more trouble with him.

With all these memories of the past, time has come and gone so fast. We wish we could live part of the past over and forget some of the past. But I believe like the old song says "Among My Souvenirs". The memories and past are things that we cherish. It seems like when you get older these memories are ever before you.

I think we should look back in life to see if we have helped someone along the way who was struggling with life's problems. If you have, I am sure you have some fond memories and life has been worth it all.

In closing, I hope you will find at least one thing to remind you of the good old days, gone but not forgotten.

I have put a few of my memories on paper mostly for my own family but whoever reads this little book I hope may receive a blessing also.

XIII

EPILOGUE

WHEN THE Campbell Lumber Mill went bankrupt in 1918 and shut down permanently, we moved to Pigeon Point. We lived directly across the Coast Road from Joe Davis and family. Joe and his wife were farming there with their seven or eight children. Their names were John, Lynn Tony, Frank, Isabel, Mary, and Evangeline Davis. There may have been one or two that I have forgotten.

The Davis family lived at Pigeon Point from the 1800s. Frank was telling me about a shingle mill that stood directly across the Coast Road from where you enter into Pigeon Point Lighthouse Station. The boiler on this mill blew up and did extensive damage to the mill. I do not know if it ever operated again after that. One of the houses that was there in 1918 was moved to where the Pheasant Farm was, and the house is still there today in 1990. The Pheasant Farm was about a mile north on the Coast Road.

When we lived at Pigeon Point in 1918, all the supplies for the lighthouse and families were brought in on a government ship every so often.

The ocean at Pigeon Point can become very treacherous at times. Joe Davis was an experienced fisherman. He went out in his rowboat for years. He understood the ocean about as well as anyone could. One day Joe and two men, hired by the Beadle Boat Company, went out to work on an anchor. They watched the ocean for about an hour before they went in their boat out to a rock where the anchor was to be repaired.

— Conant's Collection.

Pearl (Conant) Busby and Mr. Joe Davis — Pigeon Point — 1918.

The ocean was calm, no big waves all the time they watched. When they got out to the rock and climbed onto it, they had no more than got there when three monstrous waves came from nowhere. The waves covered the rock and washed them into the ocean. The two younger men got hold of Joe who was going down for the last time and held him. Joe had hip boots on which filled with water and helped pull him down. They finally got him on the rock. He swallowed so much salt water he was in bed for a week. After those three swells there was never another big wave after that. That is just the way the ocean is sometimes.

In 1917 the state hired their first highway patrolman on the coast. I believe it was the first patrolman. This was the most ridiculous thing the state could have done. There was maybe one or two cars a day. Some days there was not a single car on the Coast Highway.

I walked from Pigeon Point to the Gazos School week after week, and never saw one car. Well anyway, they hired Charley Steele and his motorcycle to patrol the rough dirt road. You could hardly get up to twenty-five miles an hour. The rocks and holes in the road were so bad you just wonder why a patrolman was needed on a road like this.

I knew Charley Steele for over sixty years. I hunted deer with him for a number of years and also with his brothers, Grover and Norman, Senior. The Steele family were real good friends of mine.

— Harvey Mowry's Collection.

Charlie Steele — First Motorcycle Policeman in San Mateo County.
Age 21 – 1917

— Harvey Mowry's Collection.

Charlie Steele — First Motorcycle Policeman in San Mateo County.
Age 21 – 1917

— Conant's Collection.

Barn in back of Pinkham's place where rum runners store champagne taken off boats near Franklin Point. Edward & father James Conant slept in the barn while on hunting trips. Barn is over 100 years old. Picture taken in 1960s.

[Facing page.]

House where Norman and Fanny Steele, Senior lived. Norman, Junior was born here in 1910. It is located on the old Steele Ranch on the Coast Highway South of the Gazos.

— Conant's Collection.

— From Ann Steele's Collection.

PESCADERO — 1913.
The flagpole can be seen at the intersection of Old Stage Road and Main Street.

[Facing page.]

PIONEERS of the PESCADERO COASTSIDE

Front (L to R) Norman Steele Sr. and son-in-law, Frank Garavanta. Back row Mr. Humphrey, Fanny Steele, Mrs. Garavanta, Charlie Steele and wife and Norman Steele Jr.

— Fanny Steele's Collection.

— From Ann Steele's Collection.

PESCADERO CREEK BRIDGE.

[Facing page.]

PESCADERO — 1916

Hauling the log for probably the first Flag Pole which was erected in the center of the Pescadero Road and the Old Stage Road. It was in later years another Flag Pole was erected away from the center of the highway a short distance South on Old Stage Road.

— Ann Steele's Collection.

— From Ann Steele's Collection.

Loren Coburn became one of the largest property owners in San Mateo County by the 1890s. And the only access to Pebble Beach crossed his land. In the 1870s he had erected a fence and posted signs "Pebble Beach — Private Roadway." But people ignored his signs, tearing them and the fence down. They considered access to the beach their birthright.

In 1891, the fence again went up with a padlock. Soon the "Pebble Beach War" commenced with violence and court battles that ran on for years. Meanwhile in 1893, Coburn announced plans to build a 200-room, $150,000 luxury resort hotel. He proceeded with construction of "Coburn's Folly." It stood vacant for years while court battles raged on.

It wasn't until 1904, a decade after its completion, that the hotel opened, but it attracted few tourists and shortly thereafter closed forever. The hotel was later carefully torn down in an attempt to salvage the 174,000 feet of redwood used in its construction.

[Facing page.]

Pebble Beach and "Coburn's Folly".

— From Lottie Manly Green's Collection.

The Pigeon Point Light House — 1918 .

[Facing page.]

FRANKLIN POINT

Graves of two English Sailors of the ship *Sir John Franklin* — wrecked in 1860, Christmas Eve. — Picture taken in 1889.

— Obtained from Lottie Manley Green's Collection.

— From Lottie Manly Green's Collection.

The "White House" was clearly visible from the sea — a landmark for sailors traveling along the coast south of Pigeon Point. Mrs. Brown on the front steps — 1900.

ABOUT the PHOTOGRAPHS

LOTTIE MANLEY GREEN lived to be 104 years of age. During the last sixteen years of her life she was cared for by Ed and Bessie Conant. For six years she was in their home, her final years being spent in a convalescent hospital nearby. Her home was in White House Canyon south of Pigeon Point. She came to live with the Conants in 1967 and passed away in 1983. Her personal photos were returned to family members. The historic photographs in the Manley collection became part of the Conant collection. q.v. *The Saga of Lottie Manley Green* following.

HARVEY H. MOWRY has exchanged photographs with the Conants on a reciprocal basis — some of their photographs appearing in his book *The Glenwood Lumber Company in the Gazos Creek Canyon, 1905 - 1908*. (unpublished.) When Frank M. Stanger was researching for his book *Sawmills in the Redwoods, Logging on the San Francisco Peninsula, 1849 - 1967*. (San Mateo Historical Society, 1967) he interviewed Jim Conant, Ed Conant's father. Dr. Stanger was in the company of Harvey Mowry on that occasion and the friendship between the Conants and the Mowrys was established. The Mowrys live at Pioneer in Amador County.

ANN STEEL, the wife of Norman Steele, acquired the photography collection of her mother-in-law, Fanny Steele. Ann Steele lived in Pescadero for many years and now resides in a Half Moon Bay retirement home. The Conants are long-standing friends of the Steeles. Ann Steele has permitted the Conants use of her photography collection.

CURTIS and LOTTIE MIRACLE were amateur photographers and shared their photographs with Jim and Ida Conant thus those photographs became part of the Conant collection. Jim Conant gave certain photographs relating to Soquel to the Soquel Pioneer and Historical Society. The remainder became the property of his son, Ed Conant.

MRS. ALICE DALE gave the photo of the Irwin Dale Box Mill to Ed Conant.

ED and BESSIE CONANT themselves took retrospective photographs of historical sites in the 1960s.

BILL JOHNSON and the U.S. Geological Survey provided the Santa Cruz Quadrangle topographic map.

The Saga of Lottie Manley Green

By Bessie Conant

LOTTIE Manley and her brothers and sisters and mother and father arrived in the White House Canyon when Lottie was but 18 months old — it was the year 1881. As their horse and wagon climbed the last hill before arriving at their newly bought home, the horse — not being used to mountains — decided to back up, down the side of the road toward the creek. He tipped the wagon over. None of the Manleys were hurt — even the horse survived.

The old Manley house was built in 1866 when the loggers and sawmills were working.

Adrian and Lottie were younger than all of the others. As years went on Mary Manley got a job in the home of Mrs. Brown. She owned the large house at the mouth of the White House Creek. Mrs. Brown hired Mary to be her housekeeper. When Mrs. Brown traveled, Mary had Lottie come and stay with her.

The sailors at sea named this large house the "White House" as a landmark. So the creek and road running back into the canyon are named "the Whitehouse Creek" and "Whitehouse Canyon Road."

It was Mary who taught Lottie her reading, writing, and arithmetic. It was too hard to go to school. The school house was up at the Pigeon Point Lighthouse area.

Each year a group of campers came into the canyon to enjoy themselves. Lottie looked forward to seeing them. Her life was very dull. Adrian had grown up and left the canyon and Lottie was very lonesome. It was hard for a young girl to meet any young man.

Meanwhile an Englishman by the name of Charlie Green was a son of wealthy parents in England. The grandparents were in a business of "Water Marks" used on important documents. There were three brothers and two sisters who all received dividends from this business. Charlie was the only one of these who got married.

Charlie would often vacation in the United States. He had friends who lived in Texas whom he liked to visit. They had friends who lived in California. These were a family of Dutchmen who were living on the mountain up above the Manleys. That property today is owned by the Santa Clara County Girl Scouts of America Campground. Because of this Dutchman and his family living there, Lottie said, they called it "Dutch Flat."

Well, Charlie was invited to visit this family. This young good-looking playboy — Straw hat, white slacks and shirt — was a lovely sight for young Lottie Manley. Of course, it was love at first sight and an opportunity to get away from that canyon. She was 21 years old.

So they were married and took off for England in 1900. Charile had promised her that they would return often — but actually, she was saying good bye to all but Mary and Adrian.

Ed and I visited England in May of 1976. Dorothy Fiedler and husband Billie Fiedler greeted us. They showed us the two houses Lottie raised them in. They are very lovely large houses still being lived in. Dorothy said her mother had very lovely furniture. Since they were renting, we believe the houses were furnished.

Dorothy went with us to see her brother, Will Green. He refused to see us. Dorothy was quite disgusted with him. She took us to the cemetery to see where her father was buried. She said he had been a real devil. Lottie told us that one night Charlie took an axe and broke her piano all to pieces. So, I believe her life wasn't very easy. The son, Will Green, took after his father's footsteps.

Then came the second World War. In the heart of London was this Water Mark business and one of Hitler's bombs made a direct hit on the building. That was the end of the money for the Green family. Times were very bad for Lottie Green. Mrs. Green lived in Southampton, England. She was there during both wars.

Bombs landed everywhere during the second World War. All the children were ordered to leave England. So Dorothy had her children scheduled to go on a ship to safety. The ship that left England ahead of them was bombed at sea and all were lost. So Dorothy kept her children with her. Dorothy was married to Billie Fiedler. He died a short time before Dorothy died. Mrs. Green died in 1983 and a couple of years later, Billie Fiedler died and a few months later Dorothy Fiedler died. Today Dorothy's children are scattered. Peter lives in Australia. He and his wife returned to England to be with his mother before she died. Peter's children were born and raised in Australia.

Peter and Rita Fiedler are retired people. All of Lottie's family pictures we sent to Lottie's daughter. There were no pictures of Lottie's mother. Hazel Fiedler Clear and husband James are retired. They live on a small island called North Hayling Island near the Island of Wight. Bob Fiedler and his wife live in Southampton. There are two other married boys living in the Southampton area.

It was near the end of the war when Charlie Green became ill with lung cancer. He was in his forties when he died. Charlie Green's family refused to help Lottie in any way. Meanwhile, at White House Canyon, Adrian Manley had moved into the building formally built by the C.C.C.s for their office building. The big house was rented. Adrian was selling off timber and other things. You see, all the remaining land belonged to Lottie and Adrian. So he couldn't sell any land without her signature.

But Lottie said Adrian was cheating on her. He wasn't sharing everything with her. So when Charlie died and the war ended, she went to get herself re-instated as an American citizen. Adrian sold some property with her approval and put $4,000.00 into an American bank account in Lottie's name. It took quite a while to get all this done. When she was once again an American citizen, she left her daughter and son and five grandchildren — never to see them again.

She flew to New York, then returned to Oakland California to Mary's home. She moved back in the White House Canyon. She moved back into the same house to which she had arrived in 1881. She had been in England for 47 years. Soon Adrian became ill — cancer of the stomach. So he moved his bed up to the old house so Lottie could take care of him. After his operation he got along fine until in 1966 he developed cancer of the esophagus and died.

After Adrian died, Lottie came with us for six years. Then she was 94 and the doctor put her into a convalescent hospital where she lived 10 years. We visited her every Sunday. If she was ill, every day. She was two miles from our home. Ed and I had bought all of Adrian and Lottie's property before he died. Then we bought the home place from her after she moved to Mountain View.

Two years ago the grandson, Peter, and wife Rita Fielder of South Australia came to California on a wonderful tour. But I think they were very sad to see where their grandmother had returned to. Dorothy and Will Fielder are both dead. Will died a short time after Dorothy died. As for anyone in Lottie's family knowing anything more than we do, They know only what we have told them. We correspond with Hazel Clear and also Peter Fielder.

We met Lottie and Adrian in October, 1961 and she died on August 4, 1983. At her request we had her body cremated and the ashes placed in the IOOF cemetery in Santa Cruz along with her brother John's grave and close by the ashes of her father and brother Adrian.

So this ends the Saga of Lottie Manley Green.

[Overleaf.]

Lottie Manley and Charlie Green. Whitehouse Canyon — 1900. Note the size of the redwood stump.

— Lottie Manley Green Collection.

Thursday, June 11, 1987 Half Moon Bay Review Page 1B

Honest Day's Work

Woodsmen Earned 40 Cents For An Hour Of Hard Labor

By Sally Finegan
Review Staff Writer

The Rev. Ed Conant said he remembers his days as a woodsman from the late '30s to the '50s as hard but as some of the best in his life.

Conant, who recently donated a 50-year-old drag saw, a forerunner of the chain saw, to the San Mateo County Historical Museum, said that even though lumbering was slow and difficult in the days before Caterpillar trucks, when great wooden logs had to be moved via blocks and cables and one man had to grapple with 300- to 400-pound shingle bolts, it was honest work.

He would do it again, he said.

"When you made a dollar, it was yours," said the 77 year-old, who one year ago retired from his ministry at the Community Church on University Avenue in East Palo Alto.

His wife, Bessie Conant, said she agreed. Now 70, she said she enjoyed living in camps and the mills, including Johannson's shingle mill on China grade in the Santa Cruz Mountains, Cecil Gibson's shingle mill near the Saratoga summit and in the Gazos Creek area.

"It was like camping out," she said.

Writing Book

Now living in Mountain View, the Conants said they would like to finish a book about the old lumbering days.

He's writing, she's editing.

People nowadays don't understand the way the work was done and the tools that were used, they said.

"You read about the old days in lumbering and it's hard to understand what it was like," said the former woodsman, who was raised on the Coastside while his father drove a bull team through the woods.

Take the drag saw he donated to the museum.

Giving it up was difficult, he said. He had owned the drag saw for 50 years, about the same length of time he and Bessie have been married (they celebrate their 50th wedding anniversary in August).

He had bought the saw at Montgomery Ward & Co. in San Jose in 1937, he said.

Part of the Family

"It was hard parting with it," he said. "It was like losing part of the family. Gee, I sure hated getting rid of that saw."

Generally working alone, he used the 200-pound drag saw to make split posts, grape stakes and shingle bolts, he said.

Used for bucking or cutting felled logs into lengths, the drag saw was just as important to the lumbering industry in those days as the chain saw, which came into use in the early '40s, is today, he said.

The drag saw Conant used was powered by a gasoline 8-horsepower, two-cycle engine, he said. The 7- to 8-foot saw was leaned against a felled log, with the top of the saw fastening onto the log via two "dogs" or pieces of spiked metal.

The other end sat on the ground, he said. The blade on the side of the machine sawed through the log.

"You drag that around, you're doing a days work," he said.

Felling Trees

To fell a tree, woodsmen would use a crosscut saw or some other falling saw, he said.

They would know which tree to fell by following the head faller, who would mark the trees and indicate the direction he wanted the tree to fall, Conant said.

MAKING 300- 400-POUND shingle bolts with a 200-pound drag saw occupied Ed Conant, left, and his brother Vernon Conant in 1946 at Cecil Gibson's shingle mill near the Saratoga summit.

Bed For Tree

Some of the trees were so big in those days that the woodsmen would have to make a bed for the tree to fall in, because "shattered wood will bind the saw," he said.

After the woodsmen felling timber would come more woodsmen like Conant who would make shingle bolts or buck the logs into lengths to get them ready to be hauled away on the skid roads, he said.

The skid roads were roads a mile or so long reaching from the forests to the mills, he said.

They were constructed by sinking 8- to 10-inch trees, leaving a few inches above the road, every 4 or 5 feet.

Then the logs would be skidded over the roads to the mills, he said.

"It was all slow work in those days," he said. "Nothing like today."

Dangerous Work

Although woodsmen in those days did not have to worry about having a saw blade bounce off a metal spike driven into a tree by radical environmentalists, the work could be dangerous as well as wearing, he said.

Conant himself never had a bad accident as a woodsman, but other woodsmen he knew did, he said.

The worst accident he recalled occurred one day in the '40s while he was working with another woodsman for Gibson's shingle mill.

Conant recounted:

The other woodsman asked for help getting a tree that he had felled "off the stump." It had begun falling and then had gotten caught in surrounding trees.

"We put some dynamite under the tree to raise it right off the stump," Conant said. "It (the explosion) sent a chunk of wood into the middle of (the other woodsman's) forehead."

The two men were 1 $\frac{1}{2}$ miles out in the woods.

Went To Hospital

Conant stuffed a handkerchief into the hole and helped get the man to a hospital in Santa Cruz. The man survived the incident.

Conant said his uncle, also a woodsman, was killed in the woods in the late 1800s while felling timber.

He was killed by what woodsmen refer to as widow-makers — the limbs torn off trees during felling that rest precariously in upper branches of other trees.

"A little bit of a breeze knocked one down and killed him" Conant said.

His lumbering days came to an end, and Conant became a minister in 1956.

For 25 years he did rescue mission work in San Jose. For eight years he had his own church near Santa Cruz, and for 21 years he was assistant pastor at the community church in Palo Alto.

NASA Mechanic

From 1941 to 1960 he also worked as an aircraft mechanic for the National Aeronautics and Space Administration.

He and Bessie raised a son, William Conant, now 48.

But the best years were the ones when he "could pick up 500 pounds," earning 40 cents per hour and working from 10 to 12 hours per day, sometimes getting in so late that he couldn't even see the road, he said.

No one took vacations because they knew there was "always somebody waiting to get your job," he said.

"I'd rather go back to those days," he said. "Even though the work was hard, I think I'd like to go back."

"I like the woods," he added later.

Gazos People I Remember from 1916-1918:

Edna Conant (sister)
Vernon Conant (brother)
Pearl Conant (sister)
Edward Conant
Willy Gregory
Jimmy Gregory
Charmion Freshour (cousin)
George Freshour (cousin)
Mable Freshour (cousin)
Hazel Freshour (cousin)
Anderson Freshour (uncle - woodsman)
George French (uncle - millwright and head sawyer)
William Gregory (saw filer)
Mrs. Gregory
Mr. Gigi (peeler)
Mr. Kelly (cook)
Mrs. Kelly (cook)
Winnie Basham
Charley Gibson (uncle)
Grandma Gibson
Jim Conant (Father, teamster and woodsman)
Ida Conant (mother)
Mr. Gaines
Mrs. Gaines
Mr. Strom (part owner)
Mrs. Strom (part owner)
David Strom (whistle punk)
David Strom's two sisters

Gazos People I Remember from 1916-1918:

Mr. Miracle (part owner)
Mrs. Miracle (part owner)
Mr. Newcomb (cook)
Mrs. Newcomb (cook)
Harry Newcomb
Sam Irwin (pond man)
Mandy Maurica (woodsman)
Mrs. Maurica
Ida Maurica
Maggie Maurica
Mr. Duncan (part owner)
Mr. Kramer (part owner)
Mr. Frank West (woods boss)
Mrs. West and son
Gloomy Gus
Mr. Fultz (cook)
Mrs. Fultz (cook)
Wally Fultz and brother
Ray Morse (fired boiler at mill)
Sticky (bull donkey operator)
Mr. Logan (cook for rock quarry in Gazos)
Mariuch (woodsman)
George Nichols
The logger killed first day at work
The Blacksmith